Innovative Manufacturing 2017

AA008742

Topical Conference at the 2017 AIChE Spring Meeting and
13th Global Congress on Process Safety

San Antonio, Texas, USA
26 – 30 March 2017

ISBN: 978-1-5108-4136-9

Printed from e-media with permission by:

Curran Associates, Inc.
57 Morehouse Lane
Red Hook, NY 12571

Some format issues inherent in the e-media version may also appear in this print version.

Copyright© (2017) by AIChE
All rights reserved.

Printed by Curran Associates, Inc. (2017)

For permission requests, please contact AIChE
at the address below.

AIChE
120 Wall Street, FL 23
New York, NY 10005-4020

Phone: (800) 242-4363
Fax: (203) 775-5177

www.aiche.org

Additional copies of this publication are available from:

Curran Associates, Inc.
57 Morehouse Lane
Red Hook, NY 12571 USA
Phone: 845-758-0400
Fax: 845-758-2633
Email: curran@proceedings.com
Web: www.proceedings.com

TABLE OF CONTENTS

485750 Smart Manufacturing Panel Discussion

Abstract:

Smart Manufacturing enables all information about the production/manufacturing process to be available when it is needed, where it is needed, and in the form it is needed across the entire manufacturing value-chain to power smart decisions. Learn about future of production and manufacturing from Smart Manufacturing experts. Hear firsthand experiences about setting strategies, supporting technologies, and lessons from implementation.

Copyright © American Institute of Chemical Engineers. All rights reserved.

2017 AIChE®
SPRING MEETING&
13TH GLOBAL CONGRESS ON PROCESS SAFETY

March 26–30, 2017
Henry B. Gonzalez Convention Center
San Antonio, TX

17AIChE Spring Meeting, San Antonio, TX

13TH GLOBAL CONGRESS ON PROCESS SAFETY

17AIChE Spring Meeting, San Antonio, TX

13TH GLOBAL CONGRESS ON PROCESS SAFETY

Value of Modern Automation in Process Industries

Laurie R. Ben
Director, Global Business Development
Emerson Automation Solutions

2017 AIChE® SPRING MEETING & 13TH GLOBAL CONGRESS ON PROCESS SAFETY

Agenda

- Setting the Stage – Fundamentals of automation
- Why Modernize?
- Where are the Benefits? Follow the Money!
- Next Steps
- Real World Example
- Summary

Automation Definition

- Consider Automation as being made up of:
  - Measurement Devices
  - A process control system to run control strategy
  - Final control elements

Fundamental Beliefs
Why Automation Provides Value

- Automation is used to make process equipment operate at expected performance
- Most process units are constructed with a less than optimal amount of measurement & control
- Installed automation technology is not currently used to full potential
- Minor process unit equipment or piping modifications may be required along with automation to improve operational performance

17 AIChE Spring Meeting, San Antonio, TX
13TH GLOBAL CONGRESS ON PROCESS SAFETY

Strategic Importance of Process Automation

- Increasing complexity and fewer resources increases reliance on technology
- Automation is only 3-5% of original CAPEX, but a huge factor in operational success
- Process automation touches every element of the process
- Successful implementation of process automation is key to the ongoing success of the operation

2017 AIChE® SPRING MEETING & 13TH GLOBAL CONGRESS ON PROCESS SAFETY

Where Benefits Come From

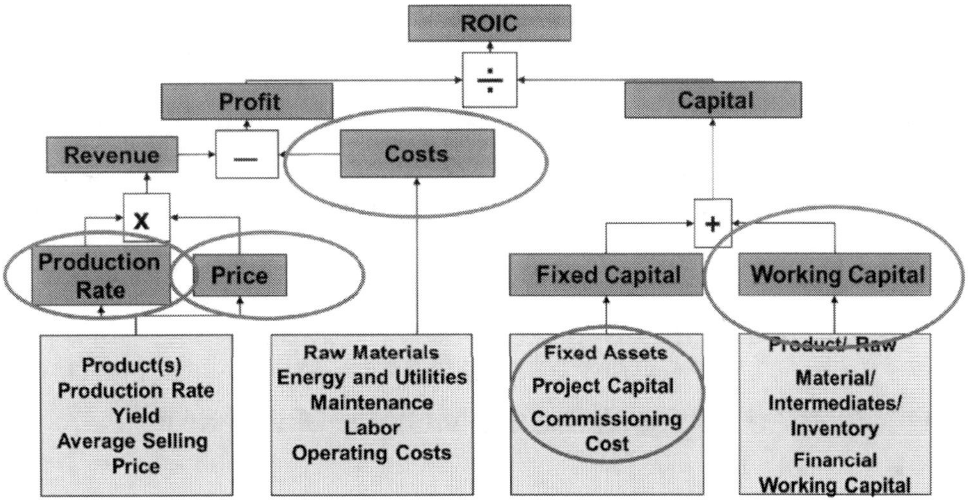

Why Modernize? Where are you on the curve?

Two types of costs associated with delaying modernization projects:

- Increasing Legacy System Health Costs
  - Parts from vendor, 3rd party, sister facilities, ebay becoming difficult, expensive, DOA,
  - Legacy expertise support from vendor, consultants, internally becoming limited
  - Site skills (operator, engineer, I&E) dwindling
- Decreasing Legacy System Automation Value (lack of competitive tools & technologies to meet business challenges)
  - Flexibility to address operational impacts/needs
  - Ability to address regulations, competitive pressures, cybersecurity, etc.
  - Expansion of process or system equipment
  - Expand employee capabilities & effectiveness

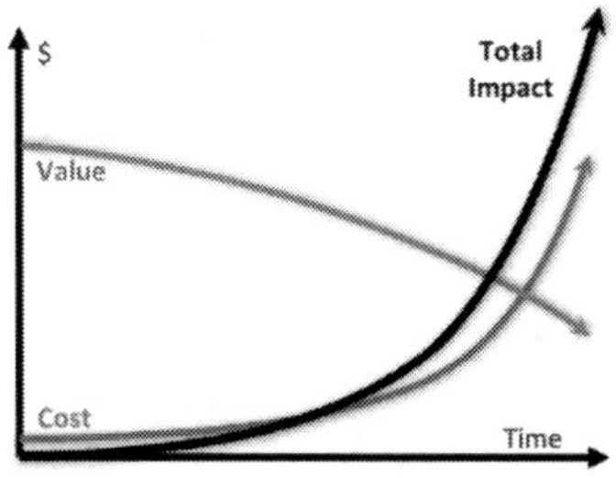

"It is not enough to just reduce the number of failures of obsolete hardware components."

"The justification must provide both cost and expected benefits (value) relative to the next best investment option."
ARC Advisory Group

2017 AIChE® SPRING MEETING & 13TH GLOBAL CONGRESS ON PROCESS SAFETY

17 AIChE Spring Meeting, San Antonio, TX — 13TH GLOBAL CONGRESS ON PROCESS SAFETY

State of the Art System Benefits

- Sustainability
  - Mature System Support Costs
  - Long Term Maintainability
- Plant & Process Unit Improvement Opportunities
- Improved Production Management
- Improved Capital Efficiency & Maintenance Spending
  - Asset Management
  - Process Automation System
- Improved Safety & Environment Incidents
- Overall Equipment Effectiveness / Improved Reliability
- Improved Logistics & Supply Chain Management

17 AIChE Spring Meeting, San Antonio, TX — 13TH GLOBAL CONGRESS ON PROCESS SAFETY

Sustaining Issues & Opportunity

- Concern of unscheduled downtime / incident do to obsolescence
- No longer cost effective to support old system
  - 30% of spare parts salvaged form older systems fail*
- Legacy system can not support new business opportunities
- Lack of capacity to expand
- Regulatory requirements
- Loss of operational intellectual talent through retirement

Legacy System MCC I/O

Modern MCC I/O

2017 AIChE® SPRING MEETING & 13TH GLOBAL CONGRESS ON PROCESS SAFETY

Plant & Process Unit Opportunities

Process Improvements

- Increased Capacity
- Production Quality
- Process Optimization

Poor Performance Causes

Variability
- Poor Control (Estimates of 2/3 of control loops under perform)
- Operators run with "safety" factory from constraint or specification
- Inconsistent batch operations
- Insufficient information for operating decisions

Deteriorating Conditions
- Equipment surface coating
- Catalyst degradation
- Equipment wear

Manual Operations

Plant Availability

17 AIChE Spring Meeting, San Antonio, TX
13TH GLOBAL CONGRESS ON PROCESS SAFETY

Increased Capacity Batch Processes

Possible Benefits

- Largest Financial Gain if Sold Out
- Excess capacity to handle large demand periods
- Emergency orders
- Alignment of production assets between made to order and made to schedule product lines

Automation Activities to Obtain

- Lower Batch Cycle times
- Ability to Effectively Monitor based on measurements vs time
- Availability of Process Equipment
  - Shared
  - In maintenance
  - Eliminate/reduce 'waiting for operator initiated actions'
- Improved scheduling of raw materials and available tankage

2017 AIChE® SPRING MEETING & 13TH GLOBAL CONGRESS ON PROCESS SAFETY

Batch Production Quality 'Right the First Time'

Possible Benefits

- Eliminates batch adjustments, major source of inconsistent cycle times
- Improved raw material yield
- Lower usage of 'processing aid' chemicals
- Reduced energy consumption
- Reduce utility usage (cooling water, chilled water / refrigerants, nitrogen)
- Frees up equipment for more production capacity
- Frees up tankage and eliminates labor & energy for rework

Automation Activities to Obtain

- Accurate raw material addition
- Improved batch reaction control, may vary
- Limit operator manual actions to reduce human error
- Electronic measurement of small manual additions with mass totals at specific time stamps

Improved Safety & Environment Incidents

Possible Benefits

- Keeping people safe
- Reduced accidents and fines
- Reduced reporting requirements and investigations
- Increased equipment availability
- Improved good will within the public community
- Reduced equipment repair cost by early notification of equipment failure

Automation Activities to Obtain

- Correct measurements of non-process control item
- Proper process interlocking including conditional interlocking
- Use of batch hold and abort functionality
- Proper design and application of alarm strategy

17 AIChE Spring Meeting, San Antonio, TX | 13TH GLOBAL CONGRESS ON PROCESS SAFETY

Automation Project Process

APPRAISE Evaluate Options	SELECT Solution Options	DEFINE Develop Solution	EXECUTE & Evaluate	OPERATE & Maintain
• FEL1/Feasibility	• FEL2/Conceptual Engineering/Pre-FEED	• FEL3/FEED	• Detailed Engineering	

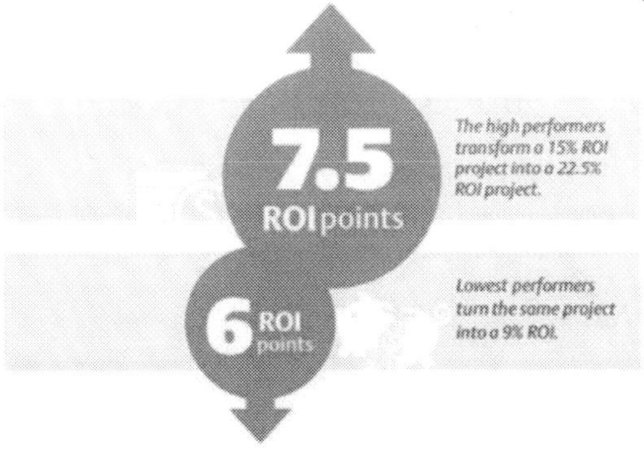

Source: Independent Project Analysis (IPA) Organization
Basis: $50M project with industry average ROI 15%

Common causes of decline

- Poor control in early planning
- Inadequate match between project and business objectives
- Paying too much for capital projects without entirely achieving business requirements

2017 AIChE® SPRING MEETING & 13TH GLOBAL CONGRESS ON PROCESS SAFETY

17 AIChE Spring Meeting, San Antonio, TX | 13TH GLOBAL CONGRESS ON PROCESS SAFETY

Design for the Justification Driver

- Determine what operational issues affect the financial KPIs
- Audit existing process for poor performance and variability
- Define opportunities for improvement through automation
- Determine what percent of the performance gap can be improved
- Estimate where the process should be running against these measures

2017 AIChE® SPRING MEETING & 13TH GLOBAL CONGRESS ON PROCESS SAFETY

Justification Identification

- Need to work with:
  - Plant Management
  - Operations
  - Planning / Scheduling / Traffic
  - Marketing and Sales
- Need to work with Accounting group to understand plant cost accounting
  - Cost allocations per unit produced
  - Activity based cost accounting
  - Sales price
  - Production costs

17 AIChE Spring Meeting, San Antonio, TX
13TH GLOBAL CONGRESS ON PROCESS SAFETY

Pre-FEED Example:
FEL 2 Leads To Successful Project

- CHALLENGES
  - High batch failure rate: 18 / 600
  - Low capacity utilization
  - Materials consumption high
  - First pass quality below targets
- STRATEGY (3-Phased Approach)
  1. Digital instrumentation & control
  2. Model Predictive Control
  3. Batch Automation
- RESULTS
  - 800+ consecutive batches
  - 33% increase in throughput
  - 7% increase in 1st pass quality
  - 56% less time spent reviewing Quality Assurance documentation
  - 80% reduction in product release documentation
  - Plans for expanded production line on hold, no need to build new process equipment units!

2017 AIChE® SPRING MEETING & 13TH GLOBAL CONGRESS ON PROCESS SAFETY
EMERSON

Pre-FEED Assessment Results - Examples

Benefits identified through automation assessment in Pre-FEED

Modernization Master Plan Is Used To Justify, Plan & Prioritize Projects

- $8 M annualized modernization benefits
- >$5M TICC estimate for 15 units
- Clarified operational procedures, issues, & future product demand
- Quantified batch cycle time variability by studying batch data
- Itemized instrumentation and controls equipment through site audit
- Prescribed an automation strategy to reduce batch cycle time variability

Modernization Opportunity Assessment Service Helps Prioritize Projects

- $30M TICC estimate: $1M budget recommended for FEED
- Advanced controls to save $500k in energy costs
- Continuous process sequential automation to improve capacity, saves >$100k energy costs & eliminates >$200k annual repair costs due to operator error
- Batch operational improvements range from $100k to over $1M

Summary

- Fundamentals of Automation
- Why Modernize – Sustaining and ROI
- Example Benefits & Automation Activities
- Best Practice of Justification Process
- Questions

480835 Automatic Control of Polymer Molecular Weight during Free Radical Polymerization

Abstract:

A model-free, fully automatic feedback molecular weight controller was developed, which uses Automatic Continuous Online Monitoring Polymerizations (ACOMP) data. ACOMP provides continuous real-time monomer and polymer concentration, weight average molecular weight and intrinsic viscosity. The model-free controller uses two proportionality constants for monomer concentration and molecular weight control; i. alpha, monomer consumption rate constant and ii. p, monomer concentration proportional to weight average kinetic chain length. The controller has been successful for molecular weight control of linear chain growth free radical polymerizations. Using the ACOMP measurements, the instantaneous weight average molecular weight can be computed and folded into a chromatography-free molecular weight distribution (MWD). Hence, the evolution of the MWD can be monitored in real time and actively controlled. Multimodal distributions were achieved through the control of monomer and chain transfer agent concentration. Chain transfer agent is an optimal control for molecular weight because it affects chain length but not reaction kinetics. Furthermore, the APMT/PolyRMC controller has successfully controlled co-monomer composition and molecular weight by computing comonomer conversion kinetics and instantaneous comonomer fraction.

Copyright © American Institute of Chemical Engineers. All rights reserved.

Automatic, Simultaneous Control of Polymer Composition and Molecular Weight during Copolymer Synthesis

Authors: Terry McAfee(1), Rick Montgomery(2), Wayne Reed(1), Thomas Zekoski(1), Aide Wu(1)

(1) Tulane University

(2) Advanced Polymer Monitoring Technologies

The ACOMP/Control Interface

Towards full feedback control of polymerization reactions

APMT Pilot System: 64 L reactor w/ 37.85 lpm fast loop

PolyRMC System: 2L reactor w/15-20 ml/min fast loop

Implemented June 2015

2

Successful Control of Monomer Concentration

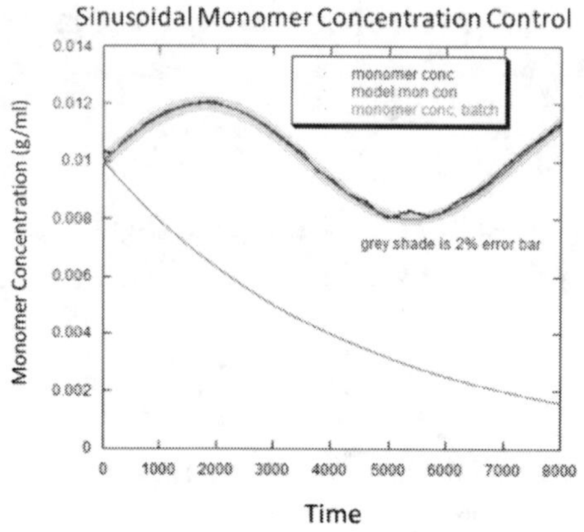

3

Successful Control of Molecular Weight

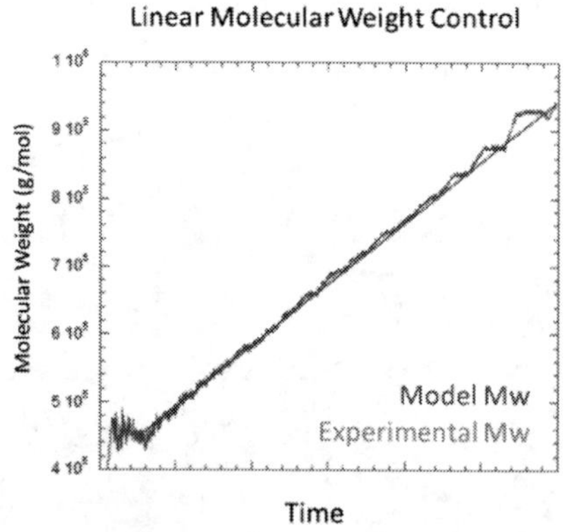

4

ACOMP's Real-time $M_{w,inst}$ holds insightful information otherwise unobtainable

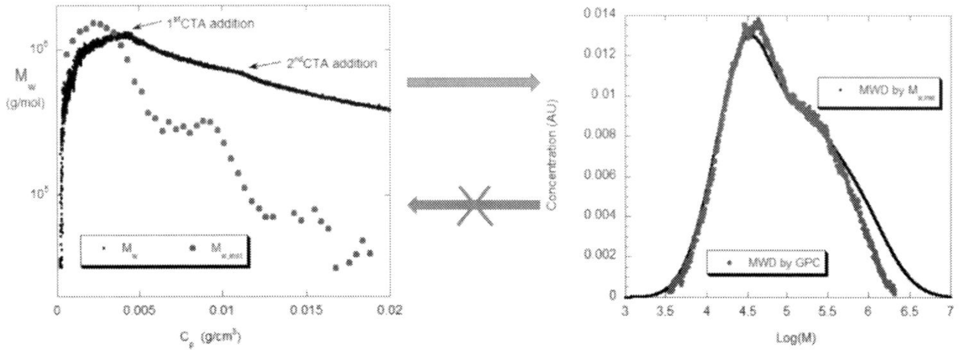

5

Terpolymerization: NMR spectra of separate monomers and combined comonomers: Styrene sulfonate, sodium Acrylate, Acrylamide

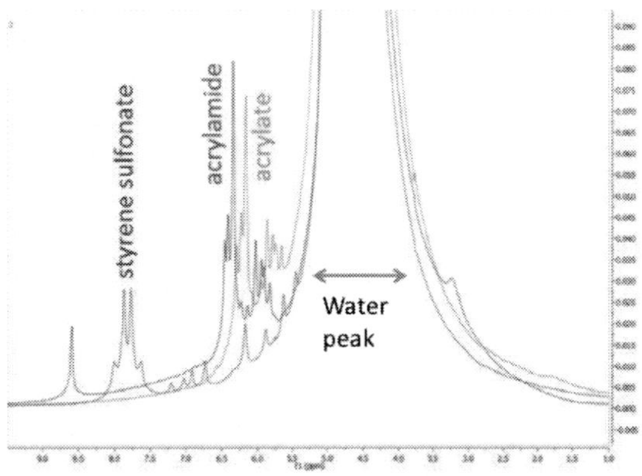

Terpolymerization data: Styrene sulfonate, sodium acrylate, acrylamide

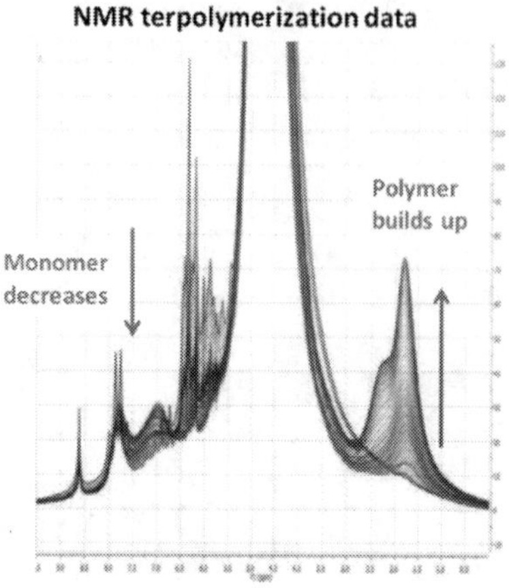

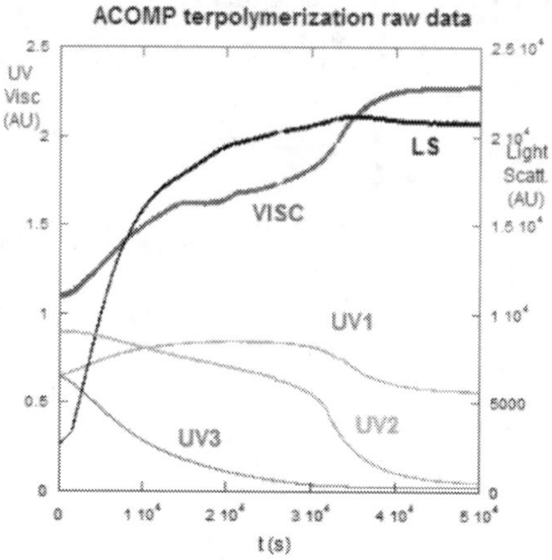

First Fully Automated Composition Controller for Copolymer Synthesis

- Does not require specific kinetic models, reactivity ratios, etc.
- Requires only knowledge of α_A and α_B, which can be directly measured throughout reactions

$$M_{w,inst} = pm \qquad m = A + B$$

$$F_A = \frac{\alpha_A A}{\alpha_A A + \alpha_B B} \qquad \Delta A_p = \alpha_A A \Delta t \quad \Delta B_p = \alpha_B B \Delta t$$

Automatically set comonomer feed pump rates to

$$Q_A(t) = \frac{V(t)}{C_A'}\left(\frac{\Delta A}{\Delta t} + \alpha_A A\right)$$

$$Q_B(t) = \frac{V(t)}{C_B'}\left(\frac{\Delta B}{\Delta t} + \alpha_B B\right)$$

Copolymerization of Acrylamide and Styrene Sulfonate

F_1 = Instantaneous Fraction of SS
C_m = Total Monomer Concentration
C_p = Total Polymer Concentration

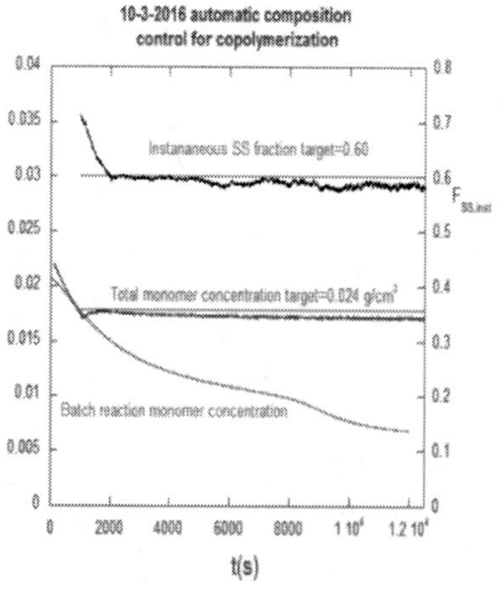

8

Simultaneous Mw and composition controller

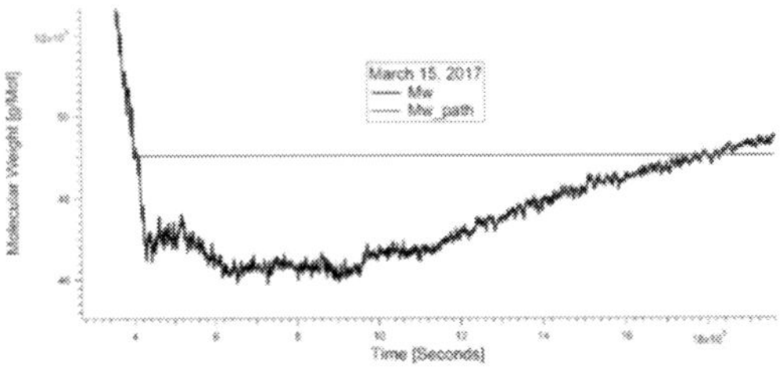

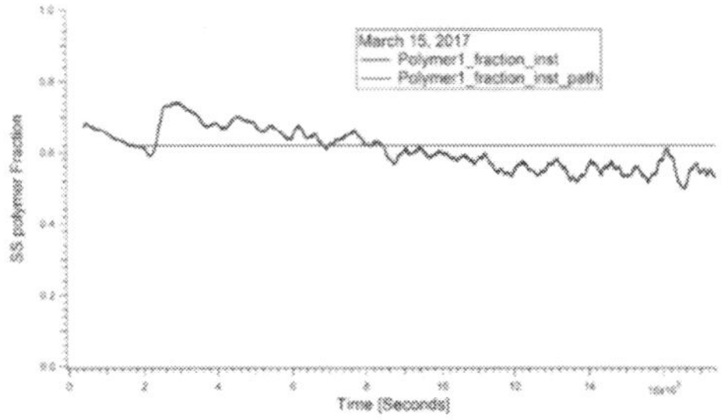

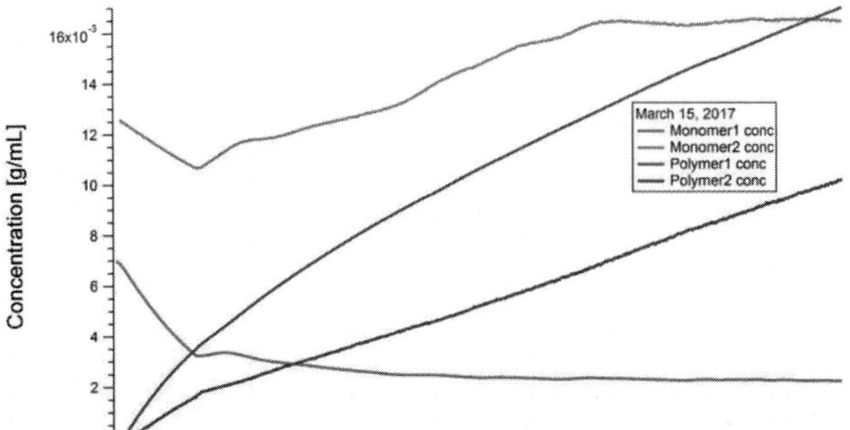

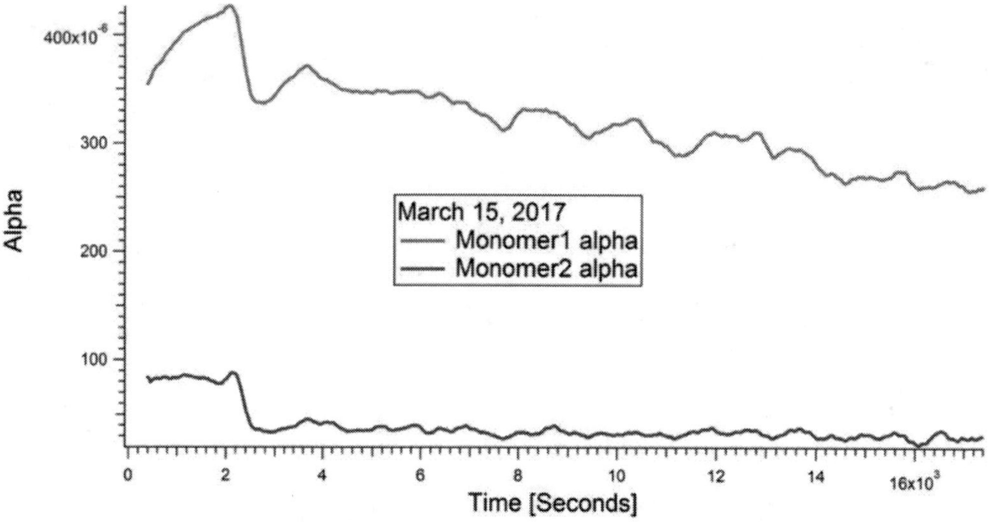

400x10⁻⁶
300
200
100
Alpha
March 15, 2017
Monomer1 alpha
Monomer2 alpha
0
2
4
6
8
10
12
14
16x10³
Time [Seconds]

Good Process Control Improves Plant Availability and Reliability

James Beall
Principal Control Consultant
Emerson Automation Solutions

Topics

- Control issues that can affect reliability
- Effective ways to deal with equipment limits
- Systems to monitor and warn operators when processes are outside desired operation window

17AIChE Spring Meeting, San Antonio, TX
13TH GLOBAL CONGRESS ON PROCESS SAFETY

How Does Poor Control Affect Equipment Reliability?

Examples:

- Pump cavitation
- Furnace coking
- Process and equipment variability
- Operating outside equipment design
- Increased shutdowns

17AIChE Spring Meeting, San Antonio, TX
13TH GLOBAL CONGRESS ON PROCESS SAFETY

Control Performance Improvements

- 4-8% increase in throughput
- 5-10% reduction in energy costs
- 2-8% reduction in product inventories
- 40-80% reduction in quality variation
- 1-5% increase in equipment availability

Pump Cavitation

- Common phenomenon with centrifugal pumps
- Caused by the formation and subsequent implosion of bubbles inside the pump casing.
- Results when Net Positive Suction Head (NPSH) is too low
- Related to the vapor pressure of the liquid at the suction conditions
- Causes physical damage to the impellers and pump casing.

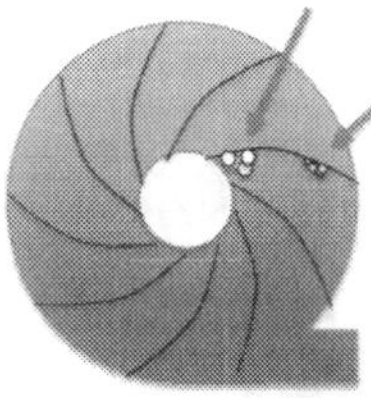

Source: http://www.pumpfundamentals.com/

What Causes Cavitation?

- Suction drum, column or tank level too low
- Inlet temperature is high
- Plugging suction strainer
- Flow too high (e.g. high line pressure drop)
- Operating parallel pumps
- Any combination of the above

17 AIChE Spring Meeting, San Antonio, TX
13TH GLOBAL CONGRESS ON PROCESS SAFETY

What Can Control Do About Cavitation?

- Stabilize level and temperature control
- Minimum level warnings
- Minimum pump suction pressure constraints
- Calculate and control minimum NPSH
- Flow balance multiple pumps
- Monitor for cavitation

17 AIChE Spring Meeting, San Antonio, TX
13TH GLOBAL CONGRESS ON PROCESS SAFETY

Furnace Coking

- Common issue for refinery furnaces
- Thermal cracking of hydrocarbon to coke which builds up on the inside of the tubes
- Reduces furnace efficiency
- Reduces heat transfer (i.e. lower outlet temp)
- Eventually requires tube decoking

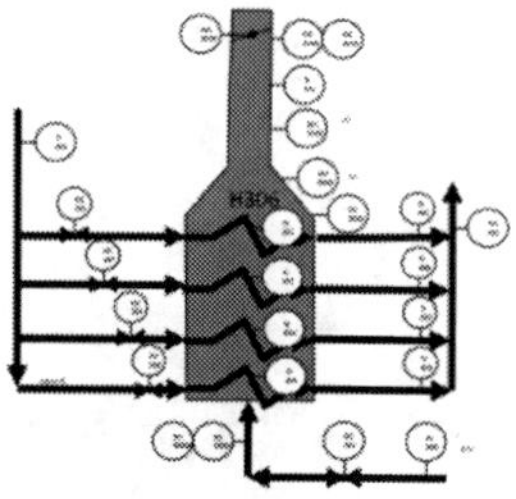

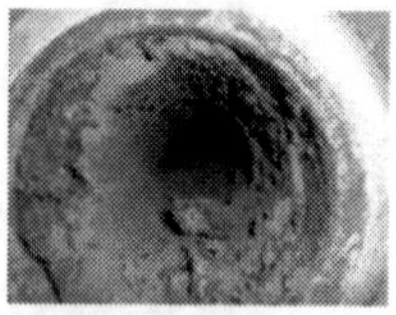

Source: http://www.asiaprotech.com/emainset.htm?id=pro03&xcode=03

2017 AIChE® SPRING MEETING & 13TH GLOBAL CONGRESS ON PROCESS SAFETY

17 AIChE Spring Meeting, San Antonio, TX — 13TH GLOBAL CONGRESS ON PROCESS SAFETY

What is Coking Affected by?

- Tube temperatures
- Hot spots
- Hydrocarbon feed composition (longer chains crack easier)
- Residence time
- Steam injection
- Flame impingement

17 AIChE Spring Meeting, San Antonio, TX — 13TH GLOBAL CONGRESS ON PROCESS SAFETY

What Can Control Do About Coking?

- Stability!!! Pass flows, fuel, steam, outlet and tube skin temperatures
- Pass balancing – Minimize pass temp differences
- Fuel BTU compensation – Use Coriolis for fuel flow measurement!
- Constraint control – Tube skin temp, pass temp differences, firebox, cross-over, tube pressure drop

17 AIChE Spring Meeting, San Antonio, TX — 13TH GLOBAL CONGRESS ON PROCESS SAFETY

Process Variability

- Control loop variability interacts throughout the process
- Oscillations affect downstream operations too
- Variability creates waste – Energy, quality, throughput, yield
- From reliability perspective, variability is hard on equipment

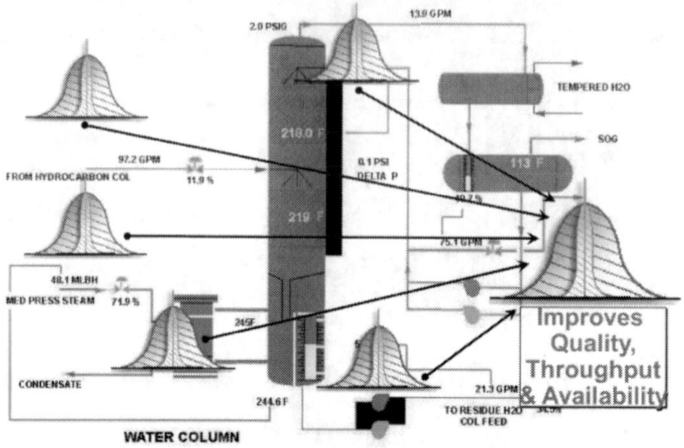

2017 AIChE® SPRING MEETING & 13TH GLOBAL CONGRESS ON PROCESS SAFETY

17 AIChE Spring Meeting, San Antonio, TX — 13TH GLOBAL CONGRESS ON PROCESS SAFETY

Variability Management

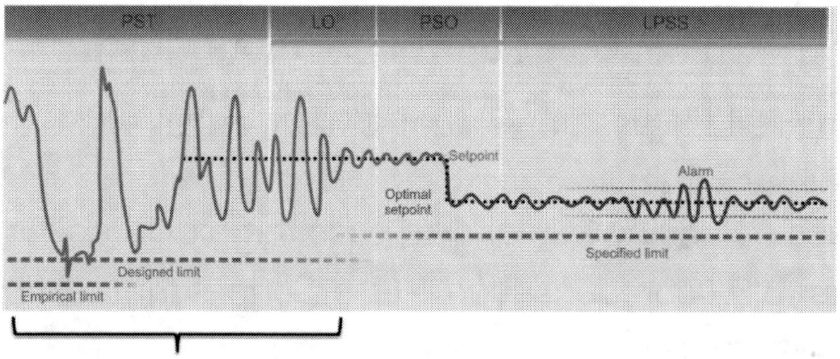

This kind of operation is hard on equipment!

2017 AIChE® SPRING MEETING & 13TH GLOBAL CONGRESS ON PROCESS SAFETY

Control Loop Monitoring Tools

- Embedded DCS tools
- Automatic Loop Tuning
- Control Loop KPI's
  - Correct mode
  - High variability
  - Oscillation index
  - Inputs bad
  - Output limited
  - Tuning index

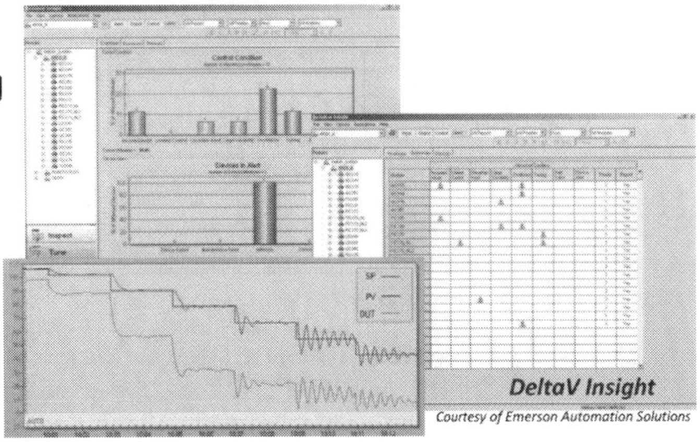

Control Performance Methodology

- Eliminate variability at the source
  - Control valves are often bad actors!
- Tune the controllers to meet control objectives
  - Most tuning methods only have "one speed"
  - Lambda tuning allows
  - Coordinate closed loop response to maximize PROCESS performance
- Utilize cascade and feed forward control
- Utilize model predictive, multivariable control (MPC)
- Minimize the affect of loop resonance

17 AIChE Spring Meeting, San Antonio, TX — 13TH GLOBAL CONGRESS ON PROCESS SAFETY

Asset Monitoring Tools

- Determine current health of equipment
- Monitor for excursions outside design operating window
- Statistically-based Analytics
- 1st Principals Models
- Notifications and alerts
- Dashboards
- Web-based implementations

17 AIChE Spring Meeting, San Antonio, TX — 13TH GLOBAL CONGRESS ON PROCESS SAFETY

Pump Failure Mode Analysis

Fault Condition	Vibration	Peak Impacting	Pressure (Discharge, Differential and Seal Fluid)			Seal Fluid Level	Temperature	Leak Detection
High Vibration	☑							
Cavitation		☑	☑					
Bearing fault		☑						
Pre-cavitation			☑					
Low head			☑	☑				
Low Discharge			☑					
Seal Pressure				☑				
Low Suction				☑				
Low Flow				☑				
Strainer Fault				☑				
Seal Failure					☑			
Seal Fluid Level						☑		
Bearing temp							☑	
Liquid HC Leak								☑

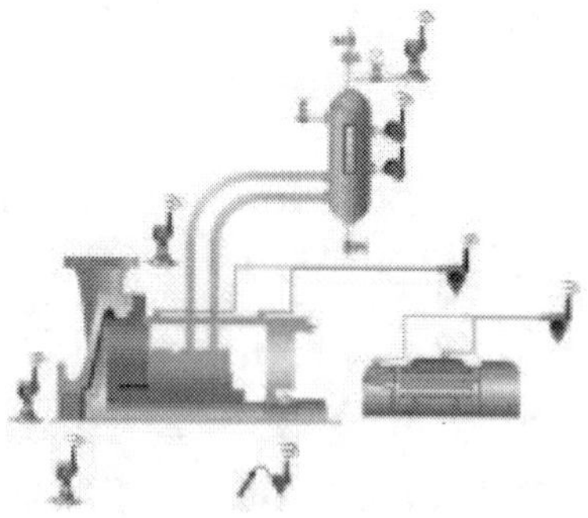

17 AIChE Spring Meeting, San Antonio, TX — 13TH GLOBAL CONGRESS ON PROCESS SAFETY

Equipment Health Indicator

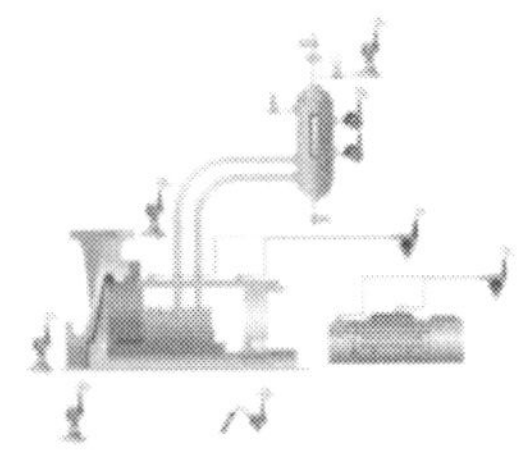

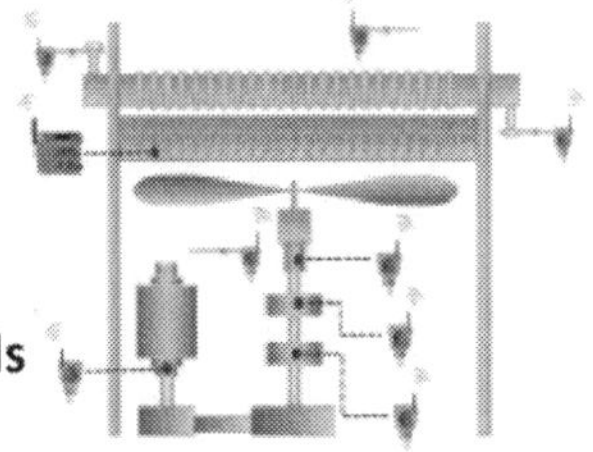

Asset Health

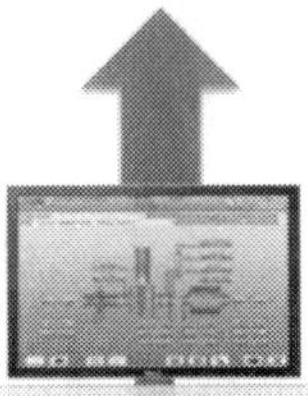

Statistical models compare actual against baseline

Process Data
Discharge/Suction P
Strainer/Filter DP
Seal Oil P/L
Flow
Air/Process T
Hydrocarbon leaks

Pre-Engineered Software Algorithm
Analyze for statistically significant changes & max/min limits

(Patented approach)

Equipment Data
Velocity/PeakVue™
Bearing T
Speed, Energy
Lvr/Pitch/Vane Pos
Run/Stop Indicator
Initial Baseline

17 AIChE Spring Meeting, San Antonio, TX — 13TH GLOBAL CONGRESS ON PROCESS SAFETY

Example Health Monitoring Dashboard

- Plant Hierarchy navigation
- Scalable across 1000's of assets
- Easy identification of issues
- Easy drill down
- Automatic notification (to the right person) of active alerts

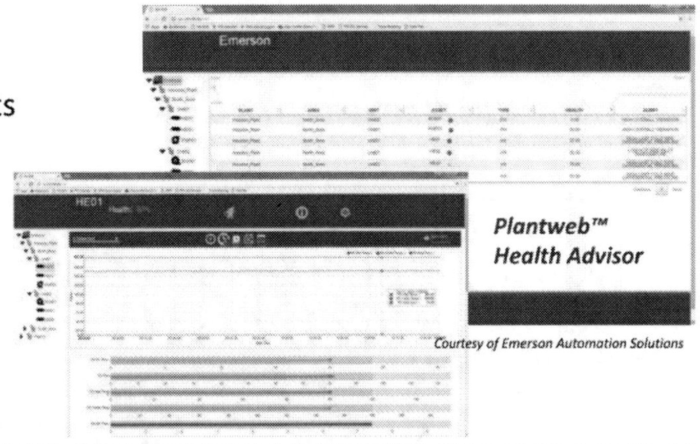

Courtesy of Emerson Automation Solutions

2017 AIChE® SPRING MEETING & 13TH GLOBAL CONGRESS ON PROCESS SAFETY

17 AIChE Spring Meeting, San Antonio, TX
13TH GLOBAL CONGRESS ON PROCESS SAFETY

Health Monitoring Allows You To...

- Identify process problems causing equipment failures
- Receive early warning on equipment degradation
- Prevent leaks, fires and emissions
- Reduce repeat failures of equipment
- Reduce unplanned maintenance costs
- Perform only required maintenance
- Quickly identify abnormal situations

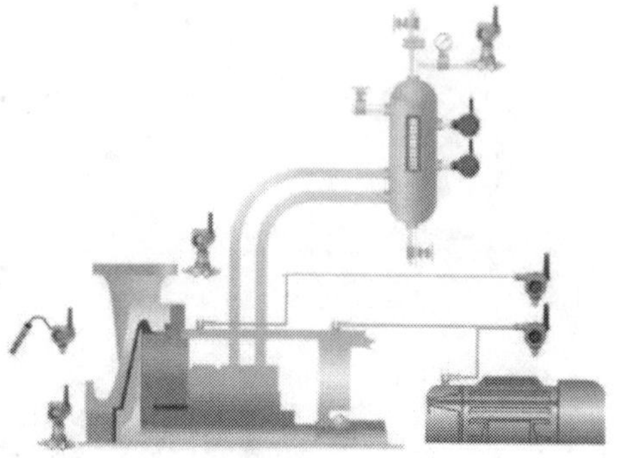

2017 AIChE® SPRING MEETING & 13TH GLOBAL CONGRESS ON PROCESS SAFETY

17 AIChE Spring Meeting, San Antonio, TX
13TH GLOBAL CONGRESS ON PROCESS SAFETY

Performance Monitoring Tools

- 1st Principal models to compare actual operation vs. design
- Key metric is efficiency & cost of degradation
- Requires equipment design curves & more measurements
- More expensive to implement
- Usually reserved for more critical equipment

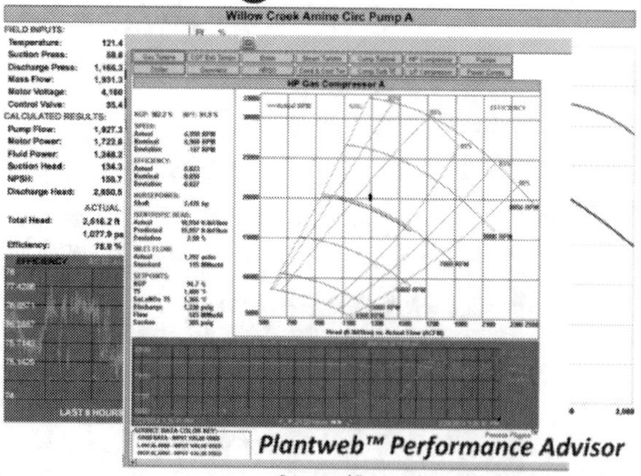

Courtesy of Emerson Automation Solutions

EMERSON

2017 AIChE® SPRING MEETING & 13TH GLOBAL CONGRESS ON PROCESS SAFETY

Performance Monitoring Allows you to…

- Get immediate notification if any monitored equipment has a fault affecting performance
- Understand degradation trends and status
- Identify potential instrument problems, pinpoint degradation sources, and evaluate the effectiveness of cost improvement actions
- **Understand financial impact of performance deviations**

The Future is in the Cloud

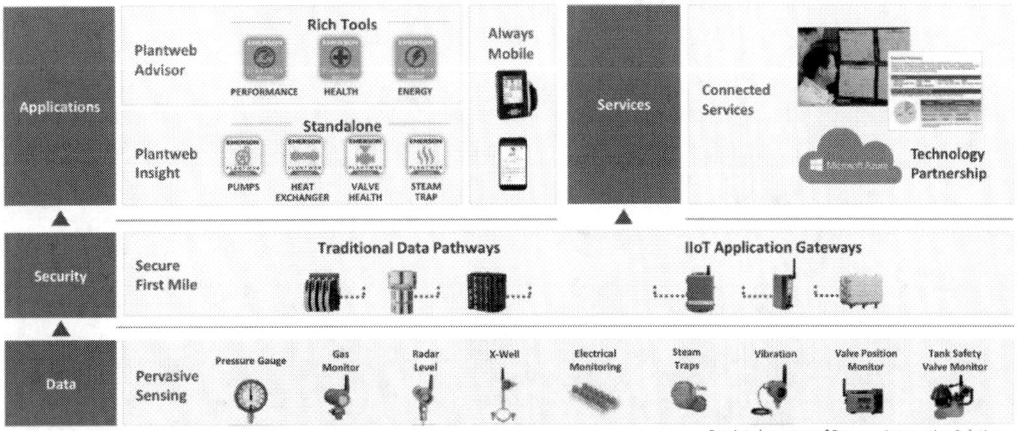

Reprinted courtesy of Emerson Automation Solutions

17AIChE Spring Meeting, San Antonio, TX
13TH GLOBAL CONGRESS ON PROCESS SAFETY

Summary

- Process control performance impacts equipment reliability/availability
  - Less variability on process and equipment
  - Less frequent shutdowns
- Asset monitoring systems
  - Identify process problems causing equipment failures
  - Receive early warning on equipment degradation
- 1-5% Improvement in availability/reliability

480961 Choosing and Using Tollers to Manufacture

Abstract:

Pilot plants are an endangered species. There are a variety of reasons for this, CAPEX limitations, safety concerns, environmental risks, resource limitations, and time constraints. CAPEX is in high demand and short supply. All projects are in competition for limited available capital funds. Regarding safety, it is key to ensure that personnel are properly trained, appropriate equipment is used, and chemical inventories are kept low. Environmentally it is important to contain and properly dispose of chemicals, which requires staff training and adherence to regulations, both federal and state. Developing qualified personnel to operate the unit requires time and training. Project cycle time is paramount and essential to maintain a competitive advantage.

The use of contract manufacturers at the pilot plant level provides a seedy solution to these problems. The majority of the investment has already been made and basic safety issues have been addressed. Only safety issues specific to the project require further assessment. Environmental risks should be discussed in detail with the contract manufacturer since you (the contracting company) own the environmental liability for the project. Lastly, the unit is already staffed with trained operators. Time constraints are now limited to toller schedule and feed logistics.

This presentation discusses

- How to identify tollers
- How to transfer the technology while maintaining IP
- How to implement the technology at the toller
- How to successfully manage the effort.

Case studies of our experiences, some of which went well, others not so well will be discussed

Copyright © American Institute of Chemical Engineers. All rights reserved.

479975 To Toll or Not to Toll: That Is the Question

Abstract:

Globalization is the result of

- the recovery of international production capacity following its devastation during the Second World War;
- the revolution in communications;
- the revolution in transportation.

The first result means that global production is at an all-time high: everybody is producing everything. The communications revolution means that everybody can market globally, cheaply via a Website on the World Wide Web. The transportation revolution means that product moves almost anywhere for almost no cost. The above bullets sum to a tremendous increase in commercial competition since the early 1990's.

One way to confront this increased competition is to reduce process and product "cycle time", where cycle time means the period between invention and commercialization. The shorter cycle time is, the faster product enters a given market. Using contract manufacturing is one method for reducing cycle time in the chemical processing industry.

This presentation discusses the "in and outs" of using contract manufacturing to reduce cycle time in the chemical processing industry. This presentation is the result of twenty-five years' experience at using contract manufacturing to reduce cycle time.

Copyright © American Institute of Chemical Engineers. All rights reserved.

AUTHOR INDEX